Discovery of the Little-Known

DISCOVERY of the LITTLE-KNOWN

Find Out What Is a Fact, a Theory, or a Conspiracy from a Christian

Paul Bryan

RESOURCE *Publications* • Eugene, Oregon

DISCOVERY OF THE LITTLE-KNOWN
Find Out What Is a Fact, a Theory, or a Conspiracy from a Christian

Resource Publications
An Imprint of Wipf and Stock Publishers
199 W. 8th Ave., Suite 3
Eugene, OR 97401

www.wipfandstock.com

PAPERBACK ISBN: 979-8-3852-8802-1
HARDCOVER ISBN: 979-8-3852-8896-0
EBOOK ISBN: 979-8-3852-8897-7

VERSION NUMBER 07/21/26

Contents

Acknowledgments

I can't express in words my appreciation for those who worked alongside me during the writing of this book. They have spent many hours reviewing, editing, and providing invaluable feedback to help make this book what it is. Their personal relationship with God and knowledge of the Bible were essential to this end. My wife, Linda, was stalwart in her assistance, reviewing every chapter.

Introduction

Construction of this Book

This book can be read or studied by individuals or groups. As you will see in this section, it was constructed to benefit a diverse audience through various learning methods. The chapters of this book include narratives, Bible verses, and questions. Some include mysteries, facts, theories, and myths about creation and evolution. These all help separate truth from fiction and enable you to know the Creator God of the Bible.

Questions to Help You Learn

Questions to help you understand and learn about a topic are included in some of the chapters.

Special Attention

Pay particular attention to the following words: Note, Important, and Definition. They provide explanations, definitions, and other information to assist in understanding the text.

Additional Books

The author has written three additional Christian non-fiction books you may find interesting. *Unraveling Creation and Evolution through Science and the Bible, Exploring Mysteries, Facts, Theories, and Myths,* will help you understand why the sciences support the biblical creation account and why they disprove various theories of evolution. If you want to discover more about the God of the Bible, I suggest our first book, *Who Is This God? A Handbook for Life with Him. Who Am I, Discovery by Personal Revelation,* will help you learn more about the person God created you to be. They can be ordered on Amazon or other online stores carrying books.

PART A

AWESOME GOD OF THE BIBLE!

1. Get to Know the Author

This Book

I usually write non-fiction books for Christian. However, by the nature of the subject matter, this book is both non-fiction and fiction.

Author's Life

A friend told me I needed to share a little about myself so readers could better connect with my writing. So, here goes.

Growing Up Years

I grew up in a small farming community. My grandparents on my mother's side were farmers. I spent much of my youth working on their farm and on others in our area. I learned a strong work ethic from these hard-working farmers.

Sheep

My grandfather was a true shepherd of his sheep. They recognized his voice and would come when he called them.

Jesus referred to Himself as our shepherd, and we are the sheep of His fold.

> I am the good shepherd; I know my sheep and my sheep know me. (John 10:14 ESV)

Gertrude the Pig

He also had pigs when he didn't have sheep. They are very smart. Gertrude was abandoned by her mother, so my grandparents put her on the porch in their home. I spent many hours feeding and playing with Gertrude. When she grew too big for the porch, they took her to the farm. When we would drive up in the pickup truck, she would come running as fast as her legs would go, oinking all the way. I was still young and very small. When she got near the fence, I would climb over the fence and get on her back. Off she would go, running all over the area with me holding on as best I could. We both thoroughly enjoyed our "joy ride."

College

I graduated with a bachelor's degree in Psychology (1969) and an MBA (1974). I found both degrees helpful in my business career.

Navy Reserves

I joined the Navy Reserve in Sacramento, California, while I was in college. I continued in active reserve duty until I graduated with my MBA. I left the reserves in 1969. As you will see later in this book, I spent a year in Japan on active duty in 1970. That's where Tammy and I lived in a small Japanese village near the tri-service communications base. It was put out of commission in 1972, and the buildings are abandoned.

Biographical Note

Paul E. Bryan's passion has been to know God and serve Him since he committed his life to Jesus Christ in 1976. Since then, he has developed and supervised ministries and taught many adult Bible study courses. This has been done for multiple churches of various denominations. The decades of ministry leadership and teaching enabled him to learn to serve people with varying degrees of spiritual maturity. This translated into the unique content and design for Who Is This God?, Unraveling Creation and Evolution, and Who Am I?

His bachelor's degree in psychology and Master's in Business Administration (MBA) were invaluable in leadership and teaching in Christian ministries and organizations, as well as in his 30-year professional career in technology management. As a university associate professor, he taught both young and mature students, which helped him hone his communication skills for different audiences. He is proficient in web design and has a proven track record for delivering high-quality content on time, meeting or exceeding publisher editorial expectations. Paul has significant skills in developing accurate, well-researched, and well-edited biblical content.

He has been a professional learning organization manager and has spent 20 years in technical roles with Electronic Data Systems. Paul has been an Associate Professor of undergraduate and graduate courses at Southern Oregon University. He worked with Campus Crusade for Christ for 5 years, implementing and managing their technology project management office. Paul is not a Bible scholar, but has taught Bible courses over multiple decades in adult Sunday School, Men's Group, and home groups. Most importantly, he knows the God of the Bible and is an active follower of Jesus Christ, seeking the wisdom and guidance of the Holy Spirit.

2. God's Names Reveal Who He Is

Names of God

God spoke to Moses out of the burning bush, and it was not consumed. Moses asked God what His name was, and He replied: "I am." Which we translate from Hebrew to Yahweh, and in Greek as Jehovah. Israel considered God's personal name of Yahweh as too sacred to speak. So they chose to use the word Jehovah, which appears nearly 7,000 times in the Bible.

> You shall not take the name of the LORD your God in vain, for the LORD will not hold him guiltless who takes his name in vain. (Exodus 20:7 ESV)

They also referred to Him as "Adonai," which means Master or Lord. In our English translation for Jehovah, we use "Lord" (small caps) or "LORD" (all CAPS). When Jehovah is used in the New Testament, it is written as "Lord."

We see another Hebrew word for God as the creator in the first chapter of Genesis. Many people assume the Creator God is the Father God. But in Colossians, we read that Jesus created everything.

> The Son is the image of the invisible God, the firstborn over all creation. *For in him all things were created: things in heaven and on earth, visible and invisible*, whether thrones or powers or rulers or authorities; all things have been created through him and for him. He is before all

> things, and in him all things hold together. (Colossians 1:15-17 NIV, author's emphasis)

Nature and Purpose of Book

This book is a compendium of topics for anyone who loves God's living word and wants to learn how to apply it more effectively in their lives. It's written to help people know the God of the Bible and how to live a life pleasing to Him.

Major Themes

The major themes in this book include the following:

- God's names
- Jesus' Apostles built the Christian world
- All this gold is merely glitter
- Experiencing Japan
- Conspiracies, theories, or facts?
- Contrary to popular belief, Columbus was not the first European to discover the Americas
- Marvels, music, and oddities
- Dragons, phoenix, and other fantastic creatures
- Founding of America, and Croatan message
- Darwin had it wrong
- Old Testament Judaism and New Testament Christianity

3. No Good News Without Jesus' Resurrection

The Good News, the Gospel, is about spending eternity with the God of the Bible. But there is no Good News without the resurrection of Jesus Christ. His death brings forgiveness of sins, but His resurrection gives eternal life.

Jesus' Resurrection Occurred According to the Promises of the Bible

The following verses confirm this for us:

> Now, brothers and sisters, I want to remind you of the gospel I preached to you, which you received and on which you have taken your stand. By this gospel you are saved, if you hold firmly to the word I preached to you. Otherwise, you have believed in vain. For what I received I passed on to you as of first importance: that Christ died for our sins according to the Scriptures, that he was buried, that he was raised on the third day according to the Scriptures, and that he appeared to Cephas, and then to the Twelve. (1 Corinthians 15:1-5 NIV)

The Gospel Is Truth, It's Not a Fable

The Gospel is the truth. There are no lies or deceptions in it. Jesus is alive and in Heaven right now, awaiting His second coming to bring all His followers into eternity and the New Earth. His words in the Gospels are words that lead us to salvation in His name.

> Salvation is found in no one else, for there is no other name under heaven given to mankind by which we must be saved. (Act 4:12 NIV)

What a wonderful Savior and Lord we have in Christ Jesus.

There have been many false theories by people trying to disprove the resurrection, such as that He fainted and was revived, or his disciples stole His body. We had a young Jewish college student doing some typing and housework. I asked her about the resurrection of Jesus, and she said that she was taught that the disciples stole His body. I showed her the scriptures about His resurrection, and she refused to believe it.

4. The Persecutor Becomes the Gospel's Best Friend

Saul came from a family of Pharisees and was one himself. Unlike many Jews in Jerusalem, he was a citizen of Jerusalem and a Roman citizen. As a Roman citizen, he has been granted privileges and access that other Jewish leaders did not have.

Persecutor of the Church

Saul was a brutal persecutor of the Christian church and believers. He would enter their homes in Jerusalem and put them in jail. As a devoted Pharisee from a family of Pharisees, he sought out Christians to stamp out "The Way," as it was initially known. The Christian church spread all the more throughout the known world as the persecuted Jews left Jerusalem to witness to other Jews.

> I persecuted the followers of this Way to their death, arresting both men and women and throwing them into prison, as the high priest and all the Council can themselves testify. I even obtained letters from them to their associates in Damascus, and went there to bring these people as prisoners to Jerusalem to be punished. (Act 22:4-5 NIV)

Saul Becomes Paul

Saul hears Jesus speak from Heaven and is converted to Christianity. The enraged Saul is pursuing Christians when he has a vision and hears Jesus speak to him from Heaven.

> "As I was on the road, approaching Damascus about noon, a very bright light from heaven suddenly shone down around me. I fell to the ground and heard a voice saying to me, 'Saul, Saul, why are you persecuting me?' "'Who are you, lord?' I asked. "And the voice replied, 'I am Jesus the Nazarene, the one you are persecuting.' The people with me saw the light but didn't understand the voice speaking to me. "I asked, 'What should I do, Lord?' "And the Lord told me, 'Get up and go into Damascus, and there you will be told everything you are to do.' "I was blinded by the intense light and had to be led by the hand to Damascus by my companions. A man named Ananias lived there. He was a godly man, deeply devoted to the law, and well regarded by all the Jews of Damascus. He came and stood beside me and said, 'Brother Saul, regain your sight.' And that very moment I could see him! "Then he told me, 'The God of our ancestors has chosen you to know his will and to see the Righteous One and hear him speak. For you are to be his witness, telling everyone what you have seen and heard. What are you waiting for? Get up and be baptized. Have your sins washed away by calling on the name of the Lord.' (Act 6-16 NLT)

Notice that Jesus sends Saul to a faithful servant of His to deliver the message of Saul's salvation.

He is later given the name Paul, but we are not told the circumstances.

> The crowd listened until Paul said that word. Then they all began to shout, "Away with such a fellow! He isn't fit to live!" (Act 22:22 NLT)

5. He Changed the World for Christ

Holy Spirit Does Miracles through Paul

Wherever the Apostle Paul goes, the Holy Spirit does miraculous signs and wonders.

> The same thing happened in Iconium. Paul and Barnabas went to the Jewish synagogue and preached with such power that a great number of both Jews and Greeks became believers. Some of the Jews, however, spurned God's message and poisoned the minds of the Gentiles against Paul and Barnabas. But the apostles stayed there a long time, preaching boldly about the grace of the Lord. *And the Lord proved their message was true by giving them power to do miraculous signs and wonders.* (Act 14:1-3 NLT, author's emphasis)

The Lord was validating His message of salvation through these signs and wonders that defied the laws of nature.

The Holy Spirit heals a man who has been lame from birth.

> While they were at Lystra, Paul and Barnabas came upon a man with crippled feet. He had been that way from birth, so he had never walked. He was sitting and listening as Paul preached. Looking straight at him, *Paul realized he had faith to be healed.* So Paul called to him in a loud voice, "Stand up!" And the man jumped

> to his feet and started walking. (Act 14:8-10 NLT, author's emphasis)

Our faith is evident when the Spirit is at work to accomplish a miracle.

Paul's Missionary Work

Without the Apostle Paul evangelizing the Jews and Gentiles, there would be little or no Christianity in the world. After he is baptized and asks Jesus to forgive him of his sins, he becomes the most prolific evangelist of his time. We do know that Peter was the first Apostle to evangelize the family and friends of the Gentile Cornelius (Act 10:1-48).

The Holy Spirit sends several partners with him on his three missionary journeys: Mark, Silas, and Barnabas. They serve in evangelism and discipleship for the Gentiles who come to Christ.

Legends about the Apostles' Demise

The New Testament book "The Acts of the Apostles" describes his missionary journeys to Jews and Gentiles.

The only Apostle in the Bible to have lived a full life is John. We have only unsubstantiated legends about how Paul and the other Apostles died.

PART B

WHAT VALUE DOES THE BIBLE ADD TO YOUR LIFE?

6. Does the Bible Add Value to Your Life?

Is your life as a believer in Jesus Christ any different if you do not study the Bible?

Canonization of the Bible

A canon is a rule or standard to go by. The biblical canon is a set of texts (also called "books") recognized by Jewish and Christian scholars as legitimate parts of our English version of the Bible. There have been debates, and there have finally been agreements, on what should be included in the Bible.

In the following verses, we see that the entire Bible is canon and is considered to be the inspired word of God:

> All Scripture is inspired by God and is useful to teach us what is true and to make us realize what is wrong in our lives. It corrects us when we are wrong and teaches us to do what is right. God uses it to prepare and equip his people to do every good work. (2 Timothy 3:16-17 NLT)

Holy Spirit Wrote the Bible

The Holy Spirit wrote the scriptures to help people live their lives in ways that please God. Therefore, all scripture is endowed with the supernatural power of the Holy Spirit and the authority of Jesus, who is Lord of all.

In the following verse, we are told that men of God spoke the

> And we have the prophetic word more fully confirmed, to which you will do well to pay attention as to a lamp shining in a dark place, until the day dawns and the morning star rises in your hearts, knowing this first of all, that *no prophecy of Scripture comes from someone's own interpretation. For no prophecy was ever produced by the will of man, but men spoke from God as they were carried along by the Holy Spirit.* (2 Peter 1:19-21 ESV, author's emphasis)

PART C

THE JOY of EXPERIENCING GOD'S LOVE

7. Do You Know the Love Languages?

Love Languages

Author Gary Chapman has identified five ways we receive love from another person.

They are as follows:

- Words of affirmation
- Acts of service
- Giving gifts
- Quality time together
- Physical touch

Every person differs in how they receive and give love to those close to them. By practicing this, we build stronger, lasting relationships. We can identify how we feel when someone uses each of the five love languages. This enables us to determine which ones we most need to receive. We can then inform our partner of this so they can continue to offer that love language.

Note: When I attended the University of California, Davis, the Primate Center was experimenting with the effects of no physical touch on baby monkeys. It was not long before they developed abnormal behaviors.

PART D

ALL THAT GLITTERS IS GOLD

8. San Francisco and Gold

San Francisco Trolley Cars

One of the traditional experiences in San Francisco is riding the trolley cars up and down the city's steep slopes. They were so crowded with tourists that I was unable to board one while I was working there. The children are delighted with this unique experience.

San Francisco and Chinatown 1975

I worked in San Francisco with a Chinese coworker (Hong Kong-based), whom I first met in a classroom at Sacramento State University, where we were both pursuing our MBAs. We became friends and then dated for about a year.

She wanted to cook me a formal Chinese meal with all the dishes she was used to. An important dish was ox tail soup. We went to a Chinese butcher in Chinatown, where she knew she could get the ox tail. She looked at the ox tail on display and told him in Chinese that it was of inferior quality. He got angry, and they argued for some time. Finally, he went to the back and brought one that was the quality she wanted.

Back at her apartment, she prepared the multi-course meal. The first dish she wanted me to try was the ox tail. I didn't like the taste and couldn't eat it. She was very disappointed. I did enjoy the rest of the meal.

California Gold Rush of 1848

John Sutter was building a saw mill along the American River in Coloma, California, when his carpenter, James Marshall, found flakes of gold in the stream in 1848. Coloma is near present-day Sacramento. When the gold rush there peaked in 1852, more than 300,000 fortune hunters came to the area seeking their fortune. They were called "forty-niners." Many used a sluice box or simply panned for the gold along the stream. Can you imagine 300,000 miners standing side by side, panning for gold?

By the end of the gold rush, San Francisco's population had increased dramatically. The California Gold Rush peaked in 1852, and by the end of the decade, it was over. The Gold Rush is credited with hastening California's admission to the Union in 1850.

Sutter's Fort 1967

My sister and I did not have parents to provide the finances for college. God provided both of us with various ways of earning money. She was a tour guide who memorized facts about Sutter's Fort and each of its rooms. I went on one of her tours, and she was impressive with her knowledge about it. The fort was a response to the gold strike in Coloma and marked the beginning of Sacramento's establishment. It was the area's first center of commerce, offering a blacksmith shop, a bakery, living quarters for its residents, and more to support the pioneers who came seeking their fortune in gold. It's considered a living history of the area, and of Sacramento in particular.

PLACER MINING 2026

If you travel along the American River just east of Sacramento, you will see huge mounds of gravel and sand piled 20 to 30 feet high. Huge dredges were brought in to cut through the river bed and, in effect, blow the loose material along the embankment. It's a marvel of industry and determination!

PART E

EXPERIENCING THE WONDERS OF JAPAN

9. Sumo Wrestling and Other Adventures in Japan

Annual Winter's Japanese Festival 1963

My hometown was ethnically and culturally diverse. We had Mexican, Spanish, Filipino, Japanese, and Caucasian families. My best friend in high school was Japanese. I spent many dinners with his farming family. We worked together on their apricot farm, bringing field lugs of apricots to the nearby weigh station. On one occasion, the weigh master rejected our truckload of fruit. This represented several thousand dollars of income for his family. Like all farmers, they needed that income for the year. His father called the supervisor in charge of the weigh stations, and the man was moved to another where Japanese farmers did not bring fruit.

Every year, the many Japanese farmers would hold a picnic in the woods. Japanese music (Sukiyaki) played on speakers hung in the trees. People throughout our small community showed up to enjoy the food, company, and music. It was really enjoyable. I still remember it today.

Japanese Restaurant in San Francisco 1963

In 1963, during my junior year at Winters High School, my class took a school bus trip to Fisherman's Wharf. After a morning of sightseeing on the Pier and seeing the seals sunning themselves on rafts, we were hungry. About a half dozen of us decided to try Japanese food at the Japanese restaurant by the Wharf. That was my first taste of sukiyaki. I loved it! I learned to make it for Tammy and me in Sacramento when we returned in 1971.

Indoor Outhouse 1970

On one occasion, Tammy and I were touring the countryside looking at temples. As evening approached, we looked for a restaurant. There was only one in the area. It was old and deteriorating. After our meal, I needed to use a restroom. What I found was an indoor outhouse. I had to step up onto a platform, drop my pants, and do my business. As I stood up from the squat, I heard a "plop" sound. Looking down, I saw my wallet sitting half-submerged. I called the owner, and she laughed. I told her I needed my wallet to pay for our meal, and that it also contained my military ID and driver's license. She gave me a long-handled hoe that was not quite long enough to reach my wallet. I stuck my head down beneath the rim and held my breath. But I could not hold it long enough. I thought I was going to pass out or vomit from the methane gas. The long trip home was not very pleasant.

This experience reminded me that my grandparents had an outdoor outhouse on their small farm. The small, old cabin did not have electricity or running water. You had to use the outhouse. However, since we had black widow spiders, you needed to look around, including under the seat.

Sumo Wrestling and God's Protection 1970

This story, set in 1970, illustrates the Holy Spirit's protection. It is about how the Holy Spirit spoke through another person to save me from what would have likely been a broken back. I was twenty-three at the time. An injury like that would have dramatically altered the rest of my life. It is important to know that God did this before I committed my life to Jesus Christ. So why did

He do it? I think it was because the injury would have altered His plan for my life.

Tammy (my first wife) and I lived in a small fishing village near the base where I was stationed in Japan. Since I was the quarterback for our Navy base football team, I would occasionally bring a football home. When he was there, I would be greeted enthusiastically by an old man who sat on a stool across from our house. I found out later that he was the village elder. One particular afternoon in the fall of 1970, he was especially animated as he pointed to the football I was carrying. Tammy taught English to a well-to-do Japanese lady, who arrived just as I went into the house. We could see Miyoko (I don't remember her name) talking with the elder. She came into the house excited. She said, "Paulsan," you have been given a great honor. The village elder asked you to participate in their annual sumo wrestling tournament. I told her I had been to a professional sumo wrestling tournament, and there was no way I would get into the ring with these mammoth athletes! Miyoko said I would disgrace the elder if I refused. So I asked her to tell him I would talk with my commanding officer to see if he would approve it, since I believed he wouldn't. However, he did agree and said that a team of five could participate. But he warned that we had better be outstanding gentlemen!

Tournament Night

We were met the night of the tournament by a young Japanese man who spoke some English. He told us how to put on the belt ("mahashi") that sumo wrestlers wear during training and tournaments. Our mahashi was made of stiff, new canvas, yes, canvas, not the silk the pros wear. As we were standing around waiting for our time in the ring, a drunk Japanese man came up from behind "Buddha," our 350-plus-pound center on the football team. He sneered at us, reached around Buddha, grabbed his mahashi from the front, and leaned back. He lifted Buddha off the ground and carried him around. His intent, of course, was to instill fear in us. We soon were told that we would be wrestling against the winners of the tournament's weight classes. By halftime, we had lost every single match. However, we realized they were using our momentum against us. As we would try to turn them in one direction (say, to the left), they would just keep us moving in that direction and then toss us to the ground or out of the ring.

The Holy Spirit Saved Me

We finally began to win some matches. I won three in a row and had just thrown a Japanese wrestler out of the ring. I stood in the circle watching to make sure he was OK when I heard someone from our team shout, "Paul, look out behind you!" As I turned, I saw the drunken Japanese man rushing at me with his head down, pointing towards the middle of my back. With adrenaline pumping, my mind raced to protect myself. I took two steps back towards him as he rushed at me, grabbed him by the back of his mahashi, took two steps forward, and flung him as hard as I could out of the ring. He landed headfirst into a tree, which cut his head and knocked him unconscious. I didn't realize the tree was there, and I wasn't trying to hurt him. As he lay there bleeding and unconscious, I looked up at our Navy and Air Force senior officers with their wives in the spectator stands. They were not pleased with me. Finally, however, the village elder and the tournament judge came up to me in the ring. They bowed down in a humble posture, asking me to forgive the drunken man's attempt to injure me. Our interpreter told me this and said they would ban the man from participating in their fall festival. The next day, people in the village treated me as if I were a celebrity. For me, it was a relief to be walking around.

Perhaps the following scripture may help you understand God's protection:

> But you, LORD, are a shield that protects me; you are my glory and the one who restores me. (Psalms 3:3 NET)

PART F

IT'S NOT ABOUT THE EASTER BUNNY

10. Customs for Christians

Holy Week and Easter Customs

Easter, like Christmas, has undergone many changes over time. Some of which have nothing to do with the resurrection of Jesus Christ. Of course, with the resurrection of Jesus, there is no Christianity. For example, Lenten fasts, eating lamb on Easter day, and Easter egg hunts for children have nothing to do with the resurrection of Jesus Christ.

The Easter story can be read in the following scriptures:

- Matthew 26-28
- Mark 14-16
- Luke 22-24
- John 18-20

Do you have a favorite Easter story preached or taught in your church?

Why do so many people who never go to church come on Easter Sunday and Christmas?

The week before Easter and the weekend of Easter provide many opportunities for worship and celebration.

These include the following:

- Sunday (Palm): crowds lay down palm branches and cheer Jesus' entry into Jerusalem (Matthew 21:1-11)
- Monday: Jesus angers vendors by throwing them out of the Temple (Matthew 21:12-46)
- Tuesday: Jesus addresses the hypocrisy of the Pharisees (Matthew 23-24)
- Wednesday: Judas is bribed by the Sanhedrin to betray Jesus in the Garden of Gethsemane (Matthew 26:14-26)
- Thursday (Maundy Thursday): Last Supper, where Judas leaves to betray Jesus, and Peter later denies Jesus three times before the rooster crows in the morning (Matthew 26:17-75)
- Friday (Good): Torture and Crucifixion of Jesus Christ (Matthew 27)
- Saturday: Was Jesus in the tomb on Saturday, or did He go to Heaven to see the Father before returning Sunday morning when Mary and the other women saw Him?

Note: That Saturday was a Sabbath, which is why the women had to wait until Sunday to anoint Jesus' body. Also, since Jewish days start at 6 pm the following day, that next morning would have been a Sunday.

- Sunday: Jesus rises from the dead on the first day of the week. (John 20:1-10)

Christmas Customs

Christmas is celebrated in many countries throughout the world. However, many people don't know its origin.

Origin of Christmas

This is the Christian festival celebrating the birth of Jesus. Since the early 20th century, Christmas has also been a secular family holiday, observed by Christians and non-Christians alike, devoid of Christian elements, and marked by an increasingly elaborate gift exchange. In this secular Christmas celebration, a mythical figure named Santa Claus plays the pivotal role. Christmastime, running from Advent (or earlier) through Epiphany, is a season rich with countless religious and secular traditions that bring together family, faith, and culture while emphasizing themes of hope, joy, generosity, and goodwill.

December 25 Date

The exact reason for selecting December 25 as the date of Jesus Christ's birth is unknown. There are several theories, most of which are linked to the time when the weather in Israel was warm enough for shepherds to graze their sheep in the open fields. Some theorize that it was the result of Christianizing the pagan "day of the birth of the unconquered sun."

Origin of Santa Claus

The name Santa Claus comes from an early Turkish figure named Nicholas. He lived in Patara in the fourth century AD. He used his family's wealth to give gifts to people. He would climb on their roofs and drop coins down their chimneys. After his death, he was referred to as St. Nick.

11. Empty Tomb, He Is Risen

The tomb was empty. Jesus was not in it. He is risen! In so doing, giving a new, resurrected life to all His followers.

We see this in the following verses:

> Now from the sixth hour darkness fell upon all the land until the ninth hour. About the ninth hour Jesus cried out with a loud voice, saying, "ELI, ELI, LAMA SABACHTHANI?" that is, "MY GOD, MY GOD, WHY HAVE YOU FORSAKEN ME?" And some of those who were standing there, when they heard it, began saying, "This man is calling for Elijah." Immediately one of them ran, and taking a sponge, he filled it with sour wine and put it on a reed, and gave Him a drink. But the rest of them said, "Let us see whether Elijah will come to save Him." And Jesus cried out again with a loud voice, and yielded up His spirit. And behold, the veil of the temple was torn in two from top to bottom; and the earth shook and the rocks were split. The tombs were opened, and many bodies of the saints who had fallen asleep were raised; and coming out of the tombs after His resurrection they entered the holy city and appeared to many. (Matthew 27:45-53 NLT)

Please also notice two important proofs that Jesus rose from the dead and was alive:

> His tomb was empty, and the heavy stone sealing the entrance had been moved.
>
> Two angels stood at the entrance to the tomb and told Mary He was not there, He had risen.

The following verses say that the death of Jesus on the cross gives His followers victory over death:

> "Where, O death, is your victory? Where, O death, is your sting?" The sting of death is sin, and the power of sin is the law. But thanks be to God! He gives us the victory through our Lord Jesus Christ. Therefore, my dear brothers and sisters, stand firm. Let nothing move you. Always give yourselves fully to the work of the Lord, because you know that your labor in the Lord is not in vain. (1 Corinthians 15:55-58 NIV)

Miraculous Proofs Jesus Rose from the Dead

The following is concrete evidence that Jesus did indeed rise from the dead after His crucifixion:

- The sun went dark for three hours when Jesus was crucified

  From noon until three in the afternoon, darkness came over all the land (Matthew 27:45 NIV)

- The temple curtain separating the holy from the holy of holies inner sanctum was torn from top to bottom

  At that moment, the curtain of the temple was torn in two from top to bottom (Matthew 27:51 NIV)

- The tombs of the dead in Jerusalem were opened, and the dead were resurrected

  The earth shook, the rocks split, and the tombs broke open. The bodies of many holy people who had died were raised to life. They came out of the tombs after Jesus' resurrection, went into the holy city, and appeared to many people.

Roman Guards

The chief priests and Pharisees went to the Roman governor, Pontius Pilate, and asked that he seal the entrance to the tomb so that Jesus' followers could not steal His body and falsely claim that Jesus had risen from the dead.

We see this in the following verses:

> The next day, that is, after the day of Preparation, the chief priests and the Pharisees gathered before Pilate and said, "Sir, we remember how that impostor said, while he was still alive, 'After three days I will rise.' Therefore order the tomb to be made secure until the third day, lest his disciples go and steal him away and tell the people, 'He has risen from the dead,' and the last fraud will be worse than the first." Pilate said to them, "You have a guard of soldiers. Go, make it as secure as you can." So they went and made the tomb secure by sealing the stone and setting a guard. (Matthew 27:62-66 ESV)

A Roman guard unit consisted of sixteen highly trained military soldiers. These were hardened, battle-tested warriors, not some wimpy group of local Romans. If they failed to prevent the body of Jesus from being stolen, the punishment was to be burned alive. So, they were obviously motivated to carry out their security duty. They were to guard the tomb until after the third day, to prevent a hoax.

In 1978, my mother lived with me in Davis, California. We hired a young Jewish girl from the university there to do some typing on her book. One day, I was talking to her about the resurrection of Jesus Christ, and she said the Jewish people knew that His disciples stole His body, so He was never resurrected. There was nothing I could say to convince her otherwise.

Securing the Tomb

To prevent the disciples from stealing the body of Jesus, Pilate ordered that it be secured in two ways:

A huge round stone weighing as much as 400 pounds was rolled over the entrance to the rock cave. It would have had to be rolled uphill, making it even more impossible to move.

A Roman seal was placed on the stone, held in place by two clay packets. The impression of a Roman signet ring was made into the clay, indicating that the grave was protected by the authority of Rome. Anyone defying the authority of Rome and its emperor would be dealt with severely.

PART G

ODDITIES and DISCOVERIES

12. Who Really Discovered the Americas?

There have been conspiracy theories about the origin of America. Those who believe this say that the ten lost tribes of Israel form the basis for our existence as an independent nation.

People believe there is documentation proving that the Phoenicians and Israelites came to America before Christopher Columbus in 1492. There were Hispanics and Anglo-Saxons on the North American continent before this. The Phoenicians and Israelites sailed through the Mediterranean and across the Atlantic centuries before the birth of Christ. Some believe that the Israelites married into Indian tribes on the East coast and lost their unique identity.

Christopher Columbus Discovers the Americas 1492

He made four trips across the Atlantic from Spain in 1492, 1493, 1498, and 1502 to find a direct water route west from Europe to Asia. But he never did. Instead, he found the Americas. He found that millions of Native Americans were already living there. This was the beginning of centuries of exploration and colonization of North and South America.

He left his two brothers in the Hispaniola settlement to rebuild it with parts of his crew's ship, using hundreds of enslaved indigenous people. He was in search of the rich gold deposits they saw in the form of many golden items used by the Indians.

Note: Columbus's 1492 ships, Niña, Pinta, and Santa Maria, were dismantled to help build the new colony.

Scottish Colonization of Nova Scotia 1621

In 1621, England and Scotland were ruled by King James VI of Scotland. England had several colonies in the New World, but Scotland didn't have any. Sir William Alexander received a charter from King James, granting him permission to establish a colony in the lands lying between New England and Newfoundland. There were already New Spain, New England, New Holland, and New France. The land was initially named Nova Scotia (Latin for New Scotland). They created baronies to entice Scots to migrate to this new land. This would create new economies, bringing additional funds to Scotland. The first settlers arrived in 1622 and settled in Port Royal (modern-day Annapolis Royal). The settlers lacked the skills to forge this new venture. Many died in the first harsh winter. In 1679, William Alexander brought seventy more settlers to bolster the settlement with skilled people. However, the ongoing war with France prevented much-needed supplies from reaching the settlers, while overland attacks from French colonial troops forced many to go back to Scotland or flee to New England. The colony's land was returned to the French, and many Scots returned to Nova Scotia, making up the majority of the settlers there. From 1770 to 1815, nearly 15,000 Scots settled in Nova Scotia. Even today, Nova Scotia has many people who identify as Scottish Canadians.

Oak Island is an island in Nova Scotia and has been the subject of many theories about pirate and Templar treasure. A TV series explores the island in search of these legendary treasures. Thousands of artifacts have been found all over the island, but no treasure to date.

Did the Vikings Discover America?

This is not as easy a question to answer as we might think. One thing is for certain: the Vikings beat Christopher Columbus to the Americas by 500 years.

Note: America was likely discovered by hunter-gatherers who crossed a land bridge from Siberia to Alaska via the Bering Strait during the Ice Age. They migrated southward along the coastline and arrived 13,000 to 35,000 years ago. Their descendants are considered the origin of Native Americans.

I had no idea about this. Did you?

Vikings Discover America

One question to consider is whether the Vikings were the first non-Native Americans to come to the Americas. Archaeologists have found evidence that the Polynesian people arrived before. But scholars are not certain when the Vikings visited North America.

Most people believe that Christopher Columbus discovered the Americas in 1492 by landing on what is today San Salvador Island, another island in the Bahamas, or the Turks and Caicos Islands. However, accounts in some Norse sagas describe Viking encounters with Vinland about 985 AD.

In 1960, a Norwegian explorer and his wife, an archaeologist, were led by a local to a site at the northern tip of Newfoundland, where they discovered the ancient remains of a Viking settlement they dated to about 1,000 AD. The Vikings were ferocious as raiders, pirates, traders, explorers, and colonizers, venturing far beyond their homeland in Scandinavia during the 9th and 11th centuries.

13. Marvels of the Aztecs and Romans

Roman Concrete Underwater 19 BC

I recently learned that the Romans developed a form of concrete suitable for underwater building projects. Over time, concrete exposed to seawater strengthens, whereas modern concrete decomposes. There is a difference between concrete and cement; cement is a binding material, while concrete is a composite made from cement and other materials. Roman concrete has been used in various ways and at various times, including the wondrous overhead aqueducts. This type of concrete was found on Oak Island in Nova Scotia.

The Aztecs Built Pyramid in Southern Mexico 400 BC

The Olmecs, Mayas, and Aztecs had their own distinct cultures, religious practices, and locations in southern Mexico. The oldest Olmec civilization is found at La Venta and dates to about 400 BC. The pyramid there stands 110 feet tall and rivals the Great Pyramid of Egypt, including its square corners.

Montezuma's Gold 1519

On November 18, 1519, the conquistador Hernando Cortez entered the Aztec capital, Tenochtitlan, now Mexico City. He overthrew Aztec control and imprisoned the Aztec ruler, Montezuma, taking control of the vast Aztec empire and the fabled mountain of gold.

Seven months later, on June 30, 1520, the Aztec people rose up against the Spaniards. However, Montezuma was stoned to death by his own people. As an angry crowd stormed his palace, Cortez and his army fled, abandoning much of his looted treasure along the way. Using torture, the Aztecs responded that the remaining gold had been "taken north." This is where fact and myth enter. Some theorists say they have found evidence of this treasure in Arizona. As far as I know, they have found some Aztec artifacts, but no gold.

There is a second myth, that the Aztec gold was hidden in a cave in a mountain in Colorado. Supposedly, an old woman named "Guatuzaca," who lived on the mountain, said the local Indians had found the silver and gold. Since the gold and silver were never found, this, too, seems to be another myth.

PART H

THE MUSIC OF YOUR LIFE

14. Music and Memories of a Lifetime

Motown Music

What about Motown music? Who is your favorite singer or group from the past?

- Diana Ross and the Supremes
- The Temptations
- Marvin Gaye
- Stevie Wonder

British Invasion

What about the "British Invasion" music? Who is your favorite singer or group from the past?

- Rolling Stones
- David Bowie
- Moody Blues
- Dusty Springfield

American Music

Then there is American music by the following artists:

- Beach Boys
- Elvis Presley
- Mamas and the Papas
- Peter, Paul, and Mary
- Lawrence Welk Show

Ed Sullivan

Ed Sullivan hosted some of the greatest entertainment of all time.

The list includes the Beatles' first appearance in America on their show.

Some other famous TV series follow:

- Roy Rogers and Dale Evans
- Lone Ranger and Tonto
- Maverick
- Have Gun Will Travel
- Hop Along Cassidy
- Howdy Doody Show
- Johnny Carson
- The Twilight Zone
- Abbot and Costello

Magazines and Games of the Past

Some long-lost magazines from the past follow:

- Saturday Evening Post
- Life

Games we enjoyed as children include the following:

- Marbles
- Pick up Sticks
- Pogo Sticks
- Erector sets

Leave Them Laughing

Some of the greatest comedians kept them laughing for decades.

The following are some memorable comedy Shows:

- Bob Hope
- Saturday Live
- George Burns and Gracie Allen Show
- Saturday Night Live
- Red Skeleton
- Rowan and Martin's Laugh-In

I think the Bob Hope Show and his support of our military overseas were my all-time favorites.

Old Sayings

There are many old sayings whose originators seem anonymous.

The following are some examples of these:

- Do you have ants in your pants?
- Holy Cow!
- It's raining cats and dogs
- Time's a wastin

PART I

ODDITIES AND MYSTERIES

15. Oddities of People, Nature, and Things

Tucker Torpedoed 1949

Preston Tucker in Ypsilanti briefly manufactured 51 Tuckers, which included the prototype. However, jealous car manufacturers created such bad press that he was forced to close production on March 3, 1949. He was proven not guilty of false allegations and acquitted. The Big Three automakers and Michigan Senator Homer S. Ferguson had a role in the Tucker Corporation's demise. Underlying the opposition was the proposed selling price of $1,00. However, those that sold did so for close to $4,000.

I remember watching the movie Tucker and thinking about the Big Three's involvement in the closure of his company.

Edsel Failure to Launch

When we lived in the State of Washington in 1958, my absentee dad brought home a new Edsel. It was a beautiful car! The front design gave it the most distinction. We were disappointed when he told us he only borrowed from a dealer friend to show us. The Ford founder, Henry Ford, launched production despite objections from Edsel's son, Henry Ford II.

San Francisco Earthquake 1906

A devastating earthquake struck San Francisco on April 18, 1906. This was a devastating earthquake that registered 7.9 on the Richter Scale. The quaking extended north and south of San Francisco along the San Andreas Fault. The shaking was felt from Los Angeles to Coos Bay, Oregon. San Francisco had experienced earthquakes in 1864, 1898, and 1900, but nothing like the one in 1906. A massive fire swept through the business area near Montgomery Street to south of Market Street. It continued for four days as firefighters struggled to extinguish it. Rain finally helped to put it completely out.

Aftermath

In the end, 28,000 buildings were destroyed by the quaking and fires. The total property loss in 1906 was estimated to be $350,000. The San Francisco yacht harbor is sinking due to debris from buildings that were pushed into it after the 1906 earthquake.

Note: I visited the San Francisco Yacht Harbor area in the 1980s. You can still see where some of the buildings were repaired from the quaking.

Northern Lights

They are also known as the "aurora borealis" and are a natural wonder. They appear as large ribbons of green and purple light. The entire sky is filled with color that is reflected off the water. They seem to magically occur as charged particles (mostly protons and electrons) from the sun collide with gases (like oxygen and nitrogen) in the Earth's upper atmosphere. They are most vividly visible near the magnetic poles. Because of increased solar activity, they can be seen more frequently and further south. They slam into Earth's upper atmosphere at speeds up to 45 million miles per hour. Earth's magnetic field protects us from the invasion.

Space Shuttle Rehab Facility Rain

Linda had a cousin who worked in the tall building where the space shuttle was rehabbed and prepped for its next mission. He said that because the building was so tall and the humidity in Titusville, Florida, was high, precipitation formed on the roof, causing rain to enter the building. The workers had to wear rain gear.

16. Oddities of Nature

Life in this natural world provides many opportunities to see unusual and surprising things and events. The following are some of these.

Your Amazing Brain

The following are twelve fascinating facts discovered through neuroscience:

- Weighs about three pounds (about 2% of the body weight)
- Regulates your autonomic system, like breathing
- Generates about 20 watts of electrical power
- Processes huge amounts of information every second
- Brain constantly rewires itself in response to experiences, learning, injury, and aging
- Brain has 86 billion neurons and 100 trillion synapses to process information
- Its neural network processes 11 million pieces of information every second
- It consumes 20% of the body's energy
- Requires a constant flow of oxygen and glucose

- It develops over a person's lifetime in 5 stages and undergoes major reorganizations at specific ages
- Targeted gene therapies are reducing disease progression
- Our memory is connected to our sleep REM cycles
- During sleep, our hippocampus replays our daily experiences, which strengthens the neural connections in the cortex up to 300%
- Brain can heal itself after injury as other regions of the brain take over the functioning of the injured area; younger brains recover more quickly
- Memories are not erased with age, even though our memory retrieval pathways may weaken, making recall more difficult, while the memories still remain stored
- Repetition helps preserve our access to memories

This knowledge of the brain is reshaping psychology, medicine, and our understanding of how it works.

Note: This is a very complicated description of the brain. One important aspect is that it can heal itself from injury (perhaps by rewiring from the injured area to a new area of the brain?).

I know several older people who had strokes, and their neurologist said they might regain some cognitive ability from their rehabilitation.

There is a myth that people can use only 10% of their brains and that training can improve higher functions.

Man in the Moon Appearance

When you look at a full moon at night, you see an odd image, which appears like the face of a man. Of course, this is a distortion of the surface of the moon. It does not have a face.

PART J

DRAGONS AND OTHER FANTASTIC CREATURES

17. Dragons and The Phoenix

Dragons

Many years ago, when Linda and I were living in Orlando, we went to Epcot Center. The Center is essentially divided into areas: The Exposition buildings and the countries. We started on the left, visiting the countries we were most interested in seeing. After many hours, we made it to China. Chinese actors were doing the Chinese dragon dances. It was fascinating to see how each person beneath the painted dragon fabric moved in sync with one another.

There have been legends of fantastic flying dragons in China for thousands of years. Are these flying serpents imaginary, or are they flying dinosaurs? I suppose the answer could be either one. Let your imagination run wild here.

Puff the Magic Dragon

"Puff the Magic Dragon" is a song about a mythical dragon by Peter, Paul, and Mary. It's a sad song as the young boy grows up and stops visiting Puff in his cave. But everyone must grow out of childhood fantasies. Some need to let go of their invisible friend they spend their time with. We all need to live in the real world in which we have been put. This is not to say that there are situations and circumstances that necessitate a change, whether moving away or leaving childhood memories behind. Life goes on with us participating as we should.

The Phoenix

The idea of a Phoenix rising out of the ashes has been around for a long time. It's shaped like a large bird with colorful plumage of red, gold, and orange. It is a symbol of rebirth, immortality, and the power of transformation. It's a mythical flying creature found in many cultures worldwide. It's seen in Greek, Roman, Egyptian, and Chinese mythologies. Its name inspired the city of Phoenix, Arizona, and the Phoenix Suns basketball team. There is a Phoenix constellation in the southern sky. It's known as the "Bennu" bird in Egyptian mythology and is associated with the pagan Ra sun god. The Chinese version is known as "Fenghuang." It's also a logo for Phoenix gasoline. It's seen across various art forms, literature, and popular culture imagery. It's also seen in books, poems, paintings, and sculptures.

Yeti ("Abominable Snowman," "Big Foot")

There is a restaurant in a town near us that has a statue of the legendary Yeti out front to draw people into it. It worked for us, as our curiosity got the best of us. Sightings of this beast have occurred all over the world, though mostly in cold climates.

Loch Ness Monster

Like the Yeti, "Nellie" has many legends surrounding its existence and even some pictures of its supposed long, snake-like body protruding above the water of the Loch (Lake).

Biblical Leviathan

Leviathan is mentioned in the biblical book of Job.

> Can you fill its hide with harpoons or its head with fishing spears? If you lay a hand on it, you will remember the struggle and never do it again! Any hope of subduing it is false; the mere sight of it is overpowering. No one is fierce enough to rouse it. Who then is able to stand against me? (Job 41:7-10 ESV)

Some people try to explain its existence as a water-based dinosaur.

Kraken

This is a squid-like sea creature that is purported to live in the waters between Iceland and Norway. Is it a myth or just a very large squid or octopus?

PART K

WHAT DO YOU REMEMBER?

18. Do You Know About These Global Events?

If you are over 65, you likely remember some of the following. I have broken them out by category.

Global Events

There have been significant events that remind us how small we are as a human species.

Consider the scope of the impact of the following:

American Moon Landing 1969

American Neil Armstrong landed on the moon on April 21, 1969. As he set foot on the moon, he said, "That's one small step for [a] man, one giant leap for mankind." The fuzzy recording made it unclear if he said "for a man" or "for man."

Soviet Union Sputnik Satellite 1969

Launched by the Soviet Union on October 4, 1969, the Russian Sputnik 1 Satellite was the first manmade satellite to orbit the

Earth. When its batteries became depleted, it fell back through Earth's atmosphere on January 4, 1958.

Berlin Wall 1961-1989

If you are old enough, you may remember President Ronald Reagan telling Russian Nikita Sergeyevich Khrushchev (the First Secretary of the Communist Party of the Soviet Union from 1953 to 1964) to tear down the Berlin Wall that separated East and West Germany. The wall was built of cinder blocks and barbed wire.

While it remained, nearly 2.5 million East Germans escaped to West Germany. Many of these were skilled workers, professionals, and intellectuals. Their loss nearly crippled the East German economy. The East Germans responded by erecting a more formidable second wall, consisting of a series of concrete walls up to 15 feet high, topped with barbed wire and guarded by manned watchtowers equipped with machine guns, and strewn with ground mines. An opening in the wall was eventually created for East Germans to escape to West Germany. In 1989, West German citizens gathered at the opening to celebrate the freedom of East Germans coming into West Germany.

In October 1989, the hard-line East German leadership was forced from power and replaced by a democratic government as democratization spread throughout Europe. On November 9, 1989, the East German government opened its borders to West Germany, ending the separation and allowing travel in both directions.

The final victory for freedom culminated on October 3, 1990, with the reunification of East and West Germany, which established Germany as a single sovereign country. East and West Berlin were also unified to form a single Berlin.

Do you remember hearing President Reagan challenge Khrushchev over the TV to tear down the Berlin Wall?

Space Shuttle Challenger Disaster 1986

Linda and I lived in Titusville, Florida, in the late 1980s. On several occasions, we went to a beach where we could see the space shuttle launch pad and watch the launch. Linda was still working in Orlando, and I was working in Melbourne, Florida. Several coworkers had previously worked for NASA on the space shuttle program. On this occasion, we went outside to watch the launch of the Challenger. Just seconds after launch, it was obvious that something had gone terribly wrong. My colleagues told me that the solid rocket boosters were jettisoned when the external tank separated from the shuttle. As the fragmented shuttle begins its descent to Earth, you can see a trail of white. A larger part of that, which we learn later, is the crew compartment for the astronauts. They all died from the crash. It's uncertain if the explosion killed them or if they ran out of oxygen on the bottom of the Atlantic Ocean. We read in the local newspaper that the crew's remains and many fragments of the shuttle were recovered over a three-month period. Engineers put the pieces out in a warehouse to determine specifically what went wrong.

Woodstock Music Festival 1969

There were other music festivals also catering to the hippie movement. Even though they sought to be profitable, most lost money. Major hippie musicians, such as Janis Joplin and Jimi Hendrix, died of drug overdoses. As a result, hippie communes died out, and the hippie movement came to a standstill.

I remember hearing Janis Joplin sing on TV in her raspy voice back in the 1960s.

PART L

AMERICA IS 250 YEARS OLD IN 2026

19. America the Beautiful

Songs Celebrating Our Nation's Independence

Star Spangled Banner

The U.S. national anthem celebrates America's victory over the British in the War of 1812. U.S. soldiers raised a large American flag over Fort McHenry in Baltimore, Maryland, on September 14, 1812. The poem was created by poet Francis Scott Key as he witnessed the fort's bombardment.

God Bless America

God Bless America was written by Irving Berlin in 1918 during World War I. It was revised in 1938, prior to World War II, and recorded by Kate Smith as her signature song. It's structured as a prayer, asking for God's blessing and guidance for our nation. One famous phrase is "stand beside her and guide her through the night."

Declaration of Independence 1776

Many Founding Fathers Said They Were a Christian

As a follower of Jesus Christ, you have the obligation to hold governmental officials accountable for their actions.

We see this in the following verses: Jesus is the foundation of our Christian heritage.

> For to us a child is born, to us a son is given; and the government shall be upon his shoulder, and his name shall be called Wonderful Counselor, Mighty God, Everlasting Father, Prince of Peace. Of the increase of his government and of peace there will be no end, on the throne of David and over his kingdom, to establish it and to uphold it with justice and with righteousness from this time forth and forevermore. The zeal of the LORD of hosts will do this. (Isaia 9:6-7 ESV)

In the next verses, we see that Christians should submit themselves to proper government authorities:

> Let every person be subject to the governing authorities. For there is no authority except from God, and those that exist have been instituted by God. Therefore whoever resists the authorities resists what God has appointed, and those who resist will incur judgment. For rulers are not a terror to good conduct, but to bad. Would you have no fear of the one who is in authority? Then do what is good, and you will receive his approval, for he is God's servant for your good. But if you do wrong, be afraid, for he does not bear the sword in vain. For he is the servant of God, an avenger who carries out God's wrath on the wrongdoer. Therefore one must be in subjection, not only to avoid God's wrath but also for the sake of conscience. For because of this you also pay taxes, for the authorities are ministers of God, attending to this very thing. Pay to all what is owed to them: taxes to whom taxes are owed, revenue to whom revenue is owed, respect to whom respect is owed, honor to whom honor is owed. (Romans 13:1-7 ESV)

The Federalist Papers

They were a series of 85 essays written by Alexander Hamilton, James Madison, and John Jay under the pseudonym Publius." They were advocating the ratification of the U.S Constitution. They were published in New York newspapers and supported key issues such as centralized federal power and the absence of a Bill of Rights. They eventually swayed public opinion.

A disagreement between the Federalists and Anti-Federalists helped shape this Country following the Revolutionary War. They each had their own ideas about this country and its federal government structure. Eventually, they reached a compromise leading to the final draft of the Constitution.

The next step was to convince the Anti-Federalist states to ratify it, which is where the Papers were born in the first place. Congress thought that the convention in Pennsylvania was the right opportunity to suggest a few minor amendments to it. However, the delegates spent months arguing over different forms of government, its powers, and the need for a more centralized federal system. One concern of some members of the convention was that the Federalists were attempting to emulate the British form of centralized government. The Anti-Federalists were concerned that smaller states with less representation would not have sufficient votes and suffer the consequences in their states. A bound collection of the Papers is available for those interested in American history. They are referred to simply as the Federalist Papers.

The Federalist Papers and Federalist books help us today to understand the various concerns of our forefathers about the Constitution. There are still arguments today about the meanings of parts of the Constitution and the need for new amendments to the Bill of Rights. The ideas of the Federalist Papers would not work in our modern society. That is because print media has a pervasive influence over people's opinions. However, it can only reach limited demographic targets.

Preamble to the Constitution 1784

The First Continental Congress met on October 26, 1774, to outline a list of rights belonging to the colonists and to threaten an economic boycott of British goods. Armed conflict broke out in the colonies within six months. Fifty-six colonial delegates from the thirteen colonies came, except those from Georgia. The restrictions by Britain and King George III resulted from the "Boston Tea Party." In response, the British banned all public meetings and revoked Massachusetts's self-government.

Ben Franklin wrote a public letter in London using an assumed name to make it clear how he felt about the laws and the British Parliament that made them. A meeting of the twelve colonists was held, including John Adams, Samuel Adams, Patrick Henry, Roger Sherman, John Jay, John Diclerson, Richard Henry Lee, and George Washington. Ben Franklin and Thomas Jefferson did not attend the meeting.

They met the morning of September 5 at the "Philadelphia Tavern." George Washington was elected President, and Peyton Randolph was Vice President of the Continental Congress. They debated, sometimes heatedly, and finally agreed to boycott all British goods within the twelve colonies. They also agreed to end exports to Britain, which included tobacco.

When the Continental Congress ended, the Massachusetts Provincial Congress named its reorganized militia the "Minutemen." George Washington led three brigades of Virginia militia in the colonists' war against Great Britain.

U.S. Constitution 1787

New York's Alexander Hamilton felt some states were acting more like independent sovereign countries. Until the new Constitution was ratified and signed, he called for a constitutional convention to reevaluate the nation's governing document. The Confederation Congress endorsed his proposal. On May 25, 1787, representatives from all thirteen states met in Philadelphia. The result was a completely new form of government. On September 17, 1787, delegates signed the new U.S. Constitution. Alexander Hamilton, James Madison, and John Jay led efforts to secure sufficient votes for ratification.

John Jay wrote the 85 essays known as "The Federalist Papers" to explain how the proposed new government would function. These were published nationwide to gain national support. Some states argued that the Constitution lacked critical protections against federal government overreach, such as freedom of religion, speech, and the press. They wanted to see a separation of state rights and federal rights. The amendments that became the "Bill of Rights" were ratified, also. After their ratification, federal elections were set to take place from Monday, December 15, 1788, to January 10, 1789.

The new federal government began on March 4, 1789. George Washington was elected the first President, and John Jay was elected Vice President. Congress also restructured the system of state representation. The Constitution was still evolving as multiple amendments were added. As with everything else in this, the federal government's location was determined through negotiations. "Washing D.C. is not a state with state rights. It is its own government overseen by the balance of power between the President, Congress, and the U.S. Supreme Court.

20. Original Thirteen Colonies

How many of the original thirteen colonies were there?

The 13 American colonies ordered by date of founding:

- Virginia (1607)
- Massachusetts (1620)
- New Hampshire (1623)
- New York (founded by the Netherlands in 1624; captured by Britain in 1664)
- Maryland (1634)
- Connecticut (1635)
- Rhode Island (1636)
- Delaware (1638)
- New Jersey (founded by the Netherlands in 1660; captured by Britain in 1664)
- North Carolina (established as Carolina in 1663; divided into North and South Carolina in 1712)
- South Carolina (established as Carolina in 1663; divided into North and South Carolina in 1712)
- Pennsylvania (1681)
- Georgia (1733)

Approximately 105 colonists of the Virginia Company came by ship from England to establish Britain's foothold in the "New World" on the James River. This area became the first capital of the Virginia colony. This triggered a frantic scramble by other European countries to establish territories in the "New World".

In the end, the British would emerge victorious. They wrested Delaware, New York, and New Jersey from the Dutch, checked the Spanish advance from Florida by establishing a presence in Georgia, and seized France's vast holdings in the wake of the French and Indian War.

Contemporary thinking is focused on the more publicized in America, the establishment of Jamestown (1607), Plymouth colony (1620), and the American Revolution (1775–83). This ignores the significant changes to colonial society prior to the American Revolution. The Declaration of Independence was signed in 1776.

Jefferson Bible (also known as "The Philosophy of Jesus")

He did not believe in the supernatural, so he removed these and created the new Jefferson Bible. He rearranged the four Gospel accounts of the life and ministry of Jesus to eliminate verses with the supernatural. He was a rationalist, like many Enlightenment intellectuals. He subtly maintained a Deistic form of Christianity. He felt creation could be understood through science.

In his first term as President in 1803, he discussed his view of a more authentic Bible about the life of Jesus with a friend, Benjamin Rush. He tried to enlist the support of a Unitarian minister, Joseph Perry. However, Perry died before he could support Jefferson, who was then left alone to promote his ideas.

He also selected verses attributed to Jesus from two King James Bibles that supported his views and created a 46-page scrapbook that distilled his "philosophy of Jesus of Nazareth." His goal was to promote his moral ideas to the American public.

Note: Jefferson obviously failed to convince America and the other countries that read the bible of his radical and misaligned ideas of social conduct and morality.

Louisiana Purchase

Louisiana Purchase, western half of the Mississippi River basin purchased in 1803 from France by the United States; at less than three cents per acre for 828,000 square miles (2,144,520 square km), it was the greatest land bargain in U.S. history. The purchase doubled the size of the United States, greatly strengthened the country materially and strategically, provided a powerful impetus to westward expansion, and confirmed the doctrine of implied powers of the federal Constitution.

This was the first significant disruption to French rule over the Louisiana Territory. The first disruption came during the Seven Years' War. In 1762, France ceded Louisiana west of the Mississippi River to Spain, and in 1763, it transferred virtually all of its remaining North American possessions to Great Britain. This arrangement, however, proved temporary. French power rebounded under the subsequent military leadership of Napoleon Bonaparte, and on October 1, 1800, Napoleon induced a reluctant King Charles IV of Spain to agree, for a consideration, to cede Louisiana back to France. King Charles gave at least his verbal assent on the condition that France would never alienate the territory to a third power. With this treaty of retrocession, known as the Treaty of San Ildefonso (confirmed March 21, 1801), would go not only the growing and commercially significant port of New Orleans but also the strategic mouth of the Mississippi River.

Paris Treaty

France had received the Territory from Spain. It was sold to the United States for $11,250,000 along with assumed claims of its citizens against France in the amount of $3,750,000. Interest payments incidental to the final settlement made the total cost of the Territory to be $27,267,622.

The Lewis and Clark Expedition

President Thomas Jefferson sent a secret message to Congress on January 18, 1803, asking for $2,500 to send a U.S. officer and a dozen soldiers to explore the area from the Missouri River to the Pacific Ocean. He wanted the expedition to increase the fur trade with local Indians along the route. He also wanted to locate the mythical "Northwest Passage." Jefferson asked his secretary, Meriwether Lewis, to lead the trip. He went to Philadelphia to learn about botany, navigation by the stars, medicine, and zoology. (WOW, what an intense learning curve.) He purchased supplies and spent $20 for "Seaman," a Newfoundland dog.

After purchasing supplies and weapons at Harpers Ferry, Virginia, he selected William Clark to assist in leading the expedition. Even though the U.S. Secretary of War denied their request for co-leadership, they were determined to share the leadership and referred to each other as "captain." Clark recruited men in Kentucky for the expedition. The "George Rogers Clark Memorial Bridge" is also known locally as the Second Street Bridge, and it crosses the Ohio River between Louisville, Kentucky, and Jeffersonville, Indiana, using Highway 31.

I remember seeing Sacagawea Park in Longview, Washington, in the 1950s. She was a scout for the Lewis and Clark expedition in the Northwest and was a member of the Indiana Shoshone Indian tribe. She was invaluable while traveling thousands of miles from North Dakota to the Pacific Ocean, assisting in relations with other Indian tribes along the way. She contributed to the expedition's natural history knowledge along the way.

Note: My wife and I have used this bridge on several occasions. However, we use two newer bridges more often. The Interstate 69 bridge crosses the Ohio River between Jeffersonville, Indiana, and Louisville, Kentucky. This bridge gets much of the traffic once using the Highway 31 bridge. The newest bridge is the East Bridge, which crosses the Ohio River on the east side of Indiana, east of Louisville, Kentucky. We use the East Bridge mostly now, since most of our doctors' appointments are in this area.

21. Croatan Message on a Tree

Archaeologists have been stumped for over 500 years. New forensic DNA sequencing, genetic genealogy, and bioarchaeological research are stunning and groundbreaking.

What really happened at Roanoke, Virginia?

The Roanoke Colony was settled in 1587 by Governor John White as England's entry into the New World. White went back to England, and when he returned in 1590, he found the camp empty. Homes were torn down, and there were no signs of violence.

Croatan Message on a Tree

Instead of using traditional archaeological methods, they used genetic markers and haplogroups in living individuals. The Lost Colony DNA Project is studying this data to find potential descendants of these settlers.

The Croatan and Lumbee tribes of North Carolina have legends of having ancestors with gray eyes, light complexions, and English surnames such as Dare, Cooper, and Brown. The DNA sampling revealed traces of European Y-chromosomes intertwined with Native American mitochondrial DNA. The findings suggest evidence of the emergence of cultural and intermarriage that resulted in Indigenous societies.

For us novices in genetics, Y-chromosome DNA traces paternal ancestry, while mitochondrial DNA traces maternal ancestry. Autosomal DNA from one test participant, tracking a broader lineage, linked to dozens of potential families in 16th-century England.

Conclusion: Evidence from the Lost Colony DNA Project transforms the research assumptions from survival to hard-fought endurance. The colonists were never lost.

English Ship "Sea Venture" and Henry Hudson in 1609

On July 25, 1609, this ship was en route to the Virginia Colony when it was deliberately driven ashore during a storm in Bermuda to prevent its sinking, and a new colony was established there.

On August 28, 1609, Henry Hudson discovered Delaware Bay, Manhattan Island, and the Native Americans living there. He explores the Hudson River, which is later named after him.

Jamestown, Virginia, Settlement in 1613

Something slightly similar in the settlement at Jamestown, Virginia. Pocahontas (1596-1617) belonged to the Powhatan tribe. She was captured and held for ransom by the English colonists during fighting in 1613. During her captivity, she converted to Christianity and married tobacco farmer John Rofe in 1614 at the age of 17 or 18. They had one child, Thomas Rolfe, in 1615.

The Rolfes traveled to England by ship in 1616, where they were presented to English society as an example of a "civilized savage" to stimulate additional investment in the Jamestown colony. She became a celebrity and was treated with grace and elegance. However, as they planned their return trip to Jamestown, she died of unknown causes in Gravesend, Kent, England, where she was buried.

Conclusion: There is no happy ending for Pocahontas and John Rolfe as there was with the Lost Roanoke colony. Her story is one of unanticipated tragedy.

Mayflower compact: Pilgrims Landed on Plymouth Rock in 1620

Ever heard of the Mayflower Compact?

In December 1620, the ship Mayflower landed in Plymouth Harbor. But before they did, the passengers signed an agreement, a compact to establish the new colony. It had over 100 passengers from England and Holland. Yet only half would survive their first brutal winter.

First Thanksgiving 1620

Note: Future generations of the original colonists use genealogy to verify their descent. Perhaps if you saw the list of the original passengers' names, you would find that you are related to one of the original families?

PART M

DARWIN DIDN'T KNOW ABOUT GENETICS

22. Was Darwin Right?

The foundational premise of the theory of biological evolution is that random events in nature drive biological and genetic changes within a species over long periods, enabling it to evolve into a new species. (This is also known as macroevolution.)

Charles Darwin Did Not Know about Genetics

Charles Darwin did not know how traits were passed on to offspring. He just observed that this occurred. He did not know about genetics and how genes encode certain traits that are passed on to future generations. He also did not know about genetic mutations and how they can affect changes in physical and behavioral characteristics in species.

Note: Charles Darwin was a Naturalist who studied and theorized about the biological and physical world. His theories did not extend beyond planet Earth into the universe, other than to say the Earth was likely older than the commonly held 6,000 years.

Darwin's Theory of Evolution by Natural Selection

In 1859, Charles Darwin published his book, *On the Origin of Species.* His theory is known as the Theory of Evolution by Natural

Selection. According to his theory, natural selection explains how diverse species could evolve from a common ancestor. (This is direct descent.) He proposed that living organisms change over long periods as they adapt to their environment. Individuals who adapt well survive and produce offspring that will survive. (Later, some people referred to this as the "survival of the fittest.") As the offspring reproduce, more of the species' population will have these new traits. Once the new traits are widespread throughout the population, the species' population is said to have evolved into a new species.

Interestingly, Darwin originally believed in the Genesis account of creation orchestrated by a "designer." It wasn't until he visited the Galápagos Islands that he began to rethink his view of the natural, biological world. As he visited various islands in the chain, he began to notice what appeared to be variations of a species. It seemed to him that evolution was occurring independently on each island as the species adapted to its unique environment. He noticed this in South American countries as well. This led Darwin to conclude that new species evolve from earlier species over long periods through natural selection. *The basis of his theory is that some better-adapted variations within a species could eventually survive and reproduce in their natural habitat, becoming a new species.*

Reminder: Darwin's ideas about evolution are still scientifically unproven theories.

Missing Links

The theory of evolution requires transitional species, also known as "missing links." A missing link is a vacant placeholder to fill a gap in the evolutionary tree for a non-existent transitional species. It's a gap in the worldwide fossil record that prevents a species from being shown to have evolved into a different species over time. For example, for vertebrates to move from the water onto land and walk, there must have been an ancient sea creature that evolved lungs to breathe air and feet to walk. This transitional species is a missing link because no fossil remains have been found to support the theory. In other words, *missing links do not exist.*

PART N

UNIVERSE REFLECTS GOD'S GLORY

23. Our Universe Speaks

The Universe and Earth Reflect the Glory and Majesty of God

The universe and Earth reflect the glory and majesty of God, so no one can say he does not exist.

> For ever since the world was created, people have seen the earth and sky. Through everything God made, they can clearly see His invisible qualities—His eternal power and divine nature. So they have no excuse for not knowing God. (Romans 1:20 NLT)

The majestic and glorious nature of the universe testifies it came from the hand of the Creator God. The nature of the universe need not be explained as having been created. This is self-evident by its glory and majesty.

Jesus Keeps Our Universe Secure

Galaxies (like our Milky Way) contain billions of stars, their related solar systems, and gases and dust. Astronomers theorize that they are moving through each other at incredible speeds with synchronized motion. How is it that they don't collide and destroy each other? The answer is that Jesus keeps the entire universe secure, just as the Father desires. Even when new stars are born and old stars die, Jesus keeps everything in its proper place by his word.

> He existed before anything else, and He holds all creation together. (Colossians 1:17 NLT)

PART O

DO ALIENS BREATHE OXYGEN?

24. Why Would God Create Aliens?

When considering the possibility of extraterrestrial life, the first question I think of is, "Why would God create aliens on other planets?" "What would be his reason and purpose for their existence?"

Earth's Atmosphere Needed for Human Life

When considering how God might have created life on other planets, we should examine how Earth's atmosphere enables life.

The following are some additional scientific facts about breathable air in the Earth's atmosphere. Our atmosphere is composed primarily of gases, with minute amounts of particulate matter.

- Dry air of the Earth's atmosphere (by quantity of molecules) is composed of 78.08 percent nitrogen, 20.95 percent oxygen, 0.93 percent argon, 0.04 percent carbon dioxide, and small amounts of other trace gases
- The amount of water vapor in the air varies depending on its location: 1 percent water vapor at sea level, and .04 percent throughout the entire atmosphere, and 4 percent in tropical climate locations
- Altitude affects the gaseous composition of air, its typical temperature, and atmospheric pressure

- Gravity creates atmospheric pressure and prevents Earth's gases from being vented into space
- Air suitable for plant photosynthesis and animal respiration is found up to 7.5 miles from the Earth's surface

Note: Theoretically, oxygen-based life on other planets would require a similar atmosphere to Earth's.

Oxygen: Basis for All Life on Earth

The basis of all plant and animal life on Earth is the essential gas, Oxygen (O). We already know that the human body is comprised of 60 percent water (H_2O) and 65 percent oxygen. So, if aliens are human-like, they must have an oxygen- and water-rich natural environment on their planet. If not, the facts about Earth's atmosphere do not apply to them. Therefore, some other gases are needed to support non-human alien life. Gases toxic to oxygen-based life on Earth include chlorine, fluorine, carbon monoxide, carbon dioxide, and many others at medium to high concentrations. Is it possible that Jesus created non-human alien life on other planets that are rich in one or more of these Earthly toxic gases?

Non-Oxygen Breathing Aliens?

Could aliens exist that thrive in environments that are not water- and oxygen-rich, such as those rich in chlorine or fluorine? God's imagination and creativity are far beyond the capability of our human minds to fathom. He can do anything he chooses. Science fiction movies and some accounts of seeing aliens on Earth depict non-human aliens as small, green, and frail, with large heads, and as unable to live in our oxygen-rich environment. Maybe their skin color is depicted as green because of the non-oxygen gases within them? This is just wild speculation with no basis in truth.

Important: If intelligent, sentient beings exist on other planets of any gaseous substance and physical form, the God of the Bible designed and created them according to his purposes.

Bible and Aliens

The most challenging question about the possibility of extraterrestrial life is, "If the Triune God created aliens to be like humans on other planets, were they made in his image?"

Important: Our Almighty Triune God could design and orchestrate whatever he wants for other planets in his universe. The Bible does not tell us whether people were created to exist on other planets. If you are interested in this possibility, you can only make assumptions and speculate about what might be.

PART P

DID CHRISTIANITY ARISE FROM JUDAISM?

25. Some Christians Struggle with the Old Testament

Christian Struggle with the Old Testament

Many born-again followers of Jesus Christ struggle to understand if and how the Ten Commandments apply to them.

Jewish People Turning to Faith in Christ

Many Israeli Jews are turning to faith in God in the midst of the wars and conflicts. Messianic Jewish communities, such as Jews for Jesus, report a rise in Bible study and in professions of faith in Christ.

The Law of Moses (Torah)

In the following verses, the Greek word translated can be better rendered as "Torah." This is because the Jews used the terms "Torah" and "Pentateuch" for the first five books of the Bible. They are: Genesis, Exodus, Leviticus, Numbers, and Deuteronomy. Jewish and Christians alike credit Moses as the human author of the Torah. We know, however, that he wrote them under the influence of the Holy Spirit.

> Above all, you must understand that no prophecy of Scripture came about by the prophet's own interpretation of things. For prophecy never had its origin in the human will, but prophets, though human, spoke from God as they were carried along by the Holy Spirit. (2 Peter 1:20-21 NIV)

Law Divided into Three Parts

God divided the Mosaic Law into three parts as follows:

- Civil: How the Jewish nation should govern itself
- Ceremonial Law: Instructions on sacrifices and special observances

God tells Christians in Colossians that they are not under these laws:

> Therefore let no one pass judgment on you in questions of food and drink, or with regard to a festival or a new moon or a Sabbath. These are a shadow of the things to come, but the substance belongs to Christ. (Colossians 2:16-17 ESV)

God Doesn't Require Christians to Obey the Moral Law of Moses

It's because it tells us how to live our personal lives, treat others with respect, and love God and others. The New Testament repeats these rules about how to treat one another. When Jesus was asked what the most important law was, He replied as follows:

> And one of the scribes came up and heard them disputing with one another, and seeing that he answered them well, asked him, "Which commandment is the most important of all?" Jesus answered, "The most important is, 'Hear, O Israel: The Lord our God, the Lord is one. And you shall love the Lord your God with all your heart and with all your soul and with all your mind and with all your strength.' The second is this: 'You shall love your

> neighbor as yourself.' There is no other commandment greater than these." (Mark 12:28-31 ESV)

How much do you love the God who created you, who sustains you, and cherishes you?

Ten Commandments

The following scripture identifies the Ten Commandments:

> And God spoke all these words, saying, I am the LORD your God, who brought you out of the land of Egypt, out of the house of slavery. You shall have no other gods before me. You shall not make for yourself a carved image, or any likeness of anything that is in heaven above, or that is in the earth beneath, or that is in the water under the earth. You shall not bow down to them or serve them, for I the LORD your God am a jealous God, visiting the iniquity of the fathers on the children to the third and the fourth generation of those who hate me, but showing steadfast love to thousands of those who love me and keep my commandments. You shall not take the name of the LORD your God in vain, for the LORD will not hold him guiltless who takes his name in vain. Remember the Sabbath day, to keep it holy. Six days you shall labor, and do all your work, but the seventh day is a Sabbath to the LORD your God. On it you shall not do any work, you, or your son, or your daughter, your male servant, or your female servant, or your livestock, or the sojourner who is within your gates. For in six days the LORD made heaven and earth, the sea, and all that is in them, and rested on the seventh day. Therefore the LORD blessed the Sabbath day and made it holy. Honor your father and your mother, that your days may be long in the land that the LORD your God is giving you. You shall not murder. You shall not commit adultery. You shall not steal. You shall not bear false witness against your neighbor. You shall not covet your neighbor's house; you shall not covet your neighbor's wife, or his male servant, or his female servant, or his ox, or his donkey, or anything that is your neighbor's. (Exodus 20:1-17 ESV)

Note: Nine of the Ten Commandments are reiterated in the New Testament.

Keeping the Sabbath

Only one of the Ten Commandments is not reiterated in the New Testament.

The following verses identify this commandment:

> The apostle Paul declares, "Therefore do not let anyone judge you by what you eat or drink, or with regard to a religious festival, a New Moon celebration or a Sabbath day. These are a shadow of the things that were to come; the reality, however, is found in Christ." (Colossians 2:16-17 ESV)
>
> One man considers one day more sacred than another; another man considers every day alike. Each one should be fully convinced in his own mind. (Romans 14:5 ESV)

Important: Keeping the Sabbath is a Jewish law, not a requirement for Christians. There is freedom for Christians regarding keeping the Sabbath or not. They will not be judged if they choose not to.

Saturday versus Sunday Celebration

The Jewish people celebrate the Sabbath on Saturday, the last day of the week. Christians celebrate the resurrection of Jesus Christ on Sundays, the first day of the week.

In the following verses, the Apostle John is in the Spirit on the Lord's Day:

> I, John, your brother and fellow partaker in the tribulation and kingdom and perseverance which are in Jesus, was on the island called Patmos because of the word of God and the testimony of Jesus. I was in the Spirit on the Lord's day, and I heard behind me a loud voice like the sound of a trumpet. (Revelation 1:9-10 NASB)

Work on Sundays?

A question I have heard from some Christians is whether they should do no work on Sundays, since they believe it is a Sabbath rest. This is an incorrect understanding of the scriptures.

What are your thoughts about whether you or another believer should work on Sundays?

What about preachers and Sunday School teachers who work on Sunday? Are they violating the fourth commandment?

PART Q

DO YOU KNOW HOW TO LIVE YOUR LIFE?

26. Startling Events in the End

There are many startling events at the end of time for our planet, solar system, and people.

Last Days

The end times are not the last days. These began with Christ's first coming and end with the rapture of all believers from the earth.

> In these last days he has spoken to us in a son, whom he appointed heir of all things, and through whom he created the world. (Hebrews 1:2 NET)

> For the Lord himself will come down from heaven with a shout of command, with the voice of the archangel, and with the trumpet of God, and the dead in Christ will rise first. Then we who are alive, who are left, will be suddenly caught up together with them in the clouds to meet the Lord in the air. And so we will always be with the Lord. Therefore encourage one another with these words. (1 Thessalonians 4:16-18 NET)

Some Christians confuse the End Times with the Last Days. It's important to read and understand the Scriptures to know the truth.

End Times

Following is a list of these end-time events:

- Rapture

  And now, dear brothers and sisters, we want you to know what will happen to the believers who have died so you will not grieve like people who have no hope. For since we believe that Jesus died and was raised to life again, we also believe that when Jesus returns, God will bring back with Him the believers who have died. We tell you this directly from the Lord: We who are still living when the Lord returns will not meet Him ahead of those who have died. For the Lord Himself will come down from heaven with a commanding shout, with the voice of the archangel, and with the trumpet call of God. First, the Christians who have died will rise from their graves. Then, together with them, we who are still alive and remain on the earth will be caught up in the clouds to meet the Lord in the air. Then we will be with the Lord forever. So encourage each other with these words. (1Thessalonians 4:13-18 NLT)

  Note: Some people don't believe in a pre-tribulation rapture of Christians. They believe it happens at Christ's second coming, after the seven-year Tribulation.

- Seven-year tribulation

  Seventy weeks have been decreed for your people and your holy city, to finish the transgression, to make an end of sin, to make atonement for iniquity, to bring in everlasting righteousness, to seal up vision and prophecy and to anoint the most holy place. So you are to know and discern that from the issuing of a decree to restore and rebuild Jerusalem until Messiah the Prince there will be seven weeks and sixty-two weeks; it will be built again, with plaza and moat, even in times of distress. Then after the sixty-two weeks the Messiah will be cut off and have nothing, and the people of the prince who is to come will destroy the city and the sanctuary. And its end will come with a flood; even to the end there will be war; desolations are determined. And he will make a firm covenant with the many for one week, but in the middle of the week he will put a stop to sacrifice and grain offering; and on the wing of abominations will come one who makes desolate, even until a complete destruction, one that is decreed, is poured out on the one who makes desolate. (Daniel 9:24-27 NASB)

- Second Coming of Jesus Christ

Then I saw heaven opened and here came a white horse! The one riding it was called "Faithful" and "True," and with justice he judges and goes to war. His eyes are like a fiery flame and there are many diadem crowns on his head. He has a name written that no one knows except himself. He is dressed in clothing dipped in blood, and he is called the Word of God. The armies that are in heaven, dressed in white, clean, fine linen, were following him on white horses. From his mouth extends a sharp sword, so that with it he can strike the nations. He will rule them with an iron rod, and he stomps the winepress of the furious wrath of God, the All-Powerful. He has a name written on his clothing and on his thigh: "King of kings and Lord of lords." (Revelation 19:11-16 NET)

Note: Some people believe this is when the "rapture" occurs.

- Millennial Reign of Christ

Then I saw an angel coming down from heaven, holding in his hand the key to the bottomless pit and a great chain. And he seized the dragon, that ancient serpent, who is the devil and Satan, and bound him for a thousand years, and threw him into the pit, and shut it and sealed it over him, so that he might not deceive the nations any longer, until the thousand years were ended. After that he must be released for a little while. Then I saw thrones, and seated on them were those to whom the authority to judge was committed. Also I saw the souls of those who had been beheaded for the testimony of Jesus and for the word of God, and those who had not worshiped the beast or its image and had not received its mark on their foreheads or their hands. They came to life and reigned with Christ for a thousand years. The rest of the dead did not come to life until the thousand years were ended. This is the first resurrection. Blessed and holy is the one who shares in the first resurrection! Over such the second death has no power, but they will be priests of God and of Christ, and they will reign with him for a thousand years. (Revelation 20:1-6 ESV)

- Current Heavens and Earth Burned Up

And by the same word, the present heavens and earth have been stored up for fire. They are being kept for the day of judgment, when ungodly people will be destroyed. But you must not forget this one thing, dear friends: A day is like a thousand years to the Lord, and a thousand years is like a day. The Lord isn't really being slow about His promise, as some people think. No, He is being patient for your sake. He does not want anyone to be destroyed, but wants everyone to repent. But the day of the Lord will come as unexpectedly as a thief. Then the heavens will pass away with a terrible noise, and the very elements themselves will disappear in fire, and the earth and everything on it will be found to deserve judgment. Since everything around us is going to be destroyed like this, what holy and godly lives you should live, looking forward to the day of God and hurrying it along. On that day, He will set the heavens on fire, and the elements will melt away in the flames. (2 Peter 3:7-12 NET)

- New Heavens and Earth Created by God

  Look! I am creating new heavens and a new earth, and no one will even think about the old ones anymore. Be glad; rejoice forever in My creation! And look! I will create Jerusalem as a place of happiness. Her people will be a source of joy. (Isaiah 65:17-18 NET)

  But we are looking forward to the new heavens and new earth He has promised, a world filled with God's righteousness. (2 Peter 3:13 NET)

- Great White Throne Judgment Day

  Then I saw a great white throne and him who was seated on it. From his presence earth and sky fled away, and no place was found for them. And I saw the dead, great and small, standing before the throne, and books were opened. Then another book was opened, which is the book of life. And the dead were judged by what was written in the books, according to what they had done. And the sea gave up the dead who were in it, Death and Hades gave up the dead who were in them, and they were judged, each one of them, according to what they had done. (Revelation 20:11-13 ESV)

- Lake of Fire

  Then he will say to those on his left, "Depart from me, you cursed, into the eternal fire prepared for the devil and his angels." (Matthew 25:41 ESV)

  It is prepared for the devil and his demons. However, all people whose names are not written in the Lamb's Book of Life will go there as well for their rebellion against God (just as the devil and his demons rebelled against God)

  Then Death and Hades were thrown into the lake of fire. This is the second death, the lake of fire. And if anyone's name was

> not found written in the book of life, he was thrown into the lake of fire. (Revelation 20:14-15 ESV)

How Should Christians Live?

Since this is the way all life on earth will end, how should Christians live their lives now?

> And now, dear children, continue in him, so that when he appears, we may be confident and unashamed before him at his coming. (1 John 2:28 NET)

> Since everything will be destroyed in this way, what kind of people ought you to be? You ought to live holy and godly lives. (2 Peter 3:11 NET)

27. Eternity: Where Will You Spend It?

Sometimes, people ask, "Isn't being a good person enough to get into Heaven?" But we must ask ourselves the following questions: "How good is good enough?" and "Whose measuring stick for goodness are we using?"

Have you ever done anything wrong in your life? If so, you are not perfect and not good enough to enter Heaven.

> Your eyes are too pure to look on evil; you cannot tolerate wrongdoing. (Habakkuk 1:13 NIV)

> You therefore must be perfect, as your heavenly Father is perfect. (Matthew 5:48 ESV)

No One Can Earn Salvation

No one is perfect enough to live in God's holy presence. Every person (no matter how nice they are) has said, thought, and done things that displease God. He calls these behaviors of wrongdoing "sin." The Greek word translated *sin* is sometimes referred to as "missing the mark" of God's perfection.

> For everyone has sinned; we all fall short of God's glorious standard. (Romans 3:23 NLT)

What about the person who says, "I have always been a Christian." Can this be true? No, it's not true. Instead, the opposite is true.

> And remember that those who do not have the Spirit of Christ living in them do not belong to Him at all. (Romans 8:9 NLT)

The Holy Spirit dwells within people only when they commit their hearts and lives to Jesus as their Savior and Lord.

There Is No Hope without Jesus Christ

The following verses describe a person's spiritual condition without Jesus Christ.

> In those days you were living apart from Christ. You were excluded from citizenship among the people of Israel, and you did not know the covenant promises God had made to them. You lived in this world without God and without hope. (Ephesians 2:12 NLT)

For the sinful nature is always hostile to God. It never did obey God's laws, and it never will. That's why those who are still under the control of their sinful nature can never please God. (Romans 8:7–8 NLT)

Note: Some born-again followers of Jesus cannot state a specific date or event when they made this decision. But they know in their hearts that they have committed themselves to him as their Savior and Lord.

Your Prayer to Know God

The following prayer is the Good News. Please pray this from your heart to enter an eternal relationship with the Father, Jesus, and the Holy Spirit.

> Dear God, thank you for loving me and caring about me. Thank you for wanting me to be with you forever and for sending Jesus to be my Savior and Lord. I believe

> that Jesus died on the cross for me, that he was dead and buried, and was raised back to life by your supernatural power, according to the scriptures.
>
> Please forgive me for all my sins against you and people. I commit to live my life under your authority and rule, Jesus, as the Lord of my life. Thank you for rescuing me from the eternal consequences of living without you.
>
> Thank you, Father, for now sending your Holy Spirit to live within me to guide, help, and teach me. I know that I have eternal life with you, my God.

If you sincerely prayed the above from your heart, you are now a born-again Christian, a follower of Jesus Christ. The Holy Spirit is living within you.

Welcome to the Eternal Family of God

You now have the following relationship with each person of the Trinity. You are a:

- Child of your Father God (Galatians 3:26)
- Disciple and brother/sister of Jesus Christ (Mark 3:35)
- Temple in which the Holy Spirit lives (1 Corinthians 6:19)

All praise and glory to our great and loving Father God!

Glossary

Aliens: Purported by some theorists to inhabit other worlds similar to planet Earth. There is no scientific proof that they exist.

Angels: They were created before the universe, and Earth was made. They are supernatural beings who exist to serve God. The devil and his demons were once angels but became corrupted when they rebelled against God in Heaven. Only a few types of God's angels are identified in the Bible.

Archaeology: This science studies the material evidence of ancient and recent human cultures. There are three main types of archaeology: (1) Prehistoric archaeology: Study of human cultures that did not have writing; (2) Protohistoric archaeology: Study of human cultures that have incomplete records; (3) Historic archaeology: Study of human cultures that have well-developed historical records. Historical records can be written or oral. Artifacts (tools, weapons, pottery, and such) from these three types provide archeologists with information about the people, their society, religious beliefs, and culture.

Bible: God's living word.

Born-Again: Jesus said that everyone must be born-again to enter the Kingdom of God (John 3:3). This occurs immediately when the Holy Spirit comes to live within people when they commit their hearts and lives to Jesus Christ as Savior and Lord. When they do this, they are saved. Conversely, people who are not born again by the Holy Spirit are not saved and are not Christians.

Christmas: A Christian celebration on December 25 for the birth of Jesus Christ. Orthodox Christians celebrate it on January 7.

Note: The difference between the Christmas dates of December 25 and January 12 stems from early church decisions made at the Council of Nicaea in 325 A.D., which established a standard for calculating Easter.

Covenant: An agreement or contract that binds the parties together for a specific purpose.

Creation: The Bible describes creation in the first two chapters of the Old Testament book of Genesis. The first verse in the Bible says that the Triune God existed before creation. *In the beginning, God created the heavens and the earth* (Genesis 1:1 ESV).

Darwinian Evolution (Theory of Evolution by Natural Selection) – In 1859, Charles Darwin published his book, On the Origin of Species. His theory is known as the Theory of Evolution by Natural Selection. He proposes that living organisms change over long periods as they adapt to their environment. Individuals who adapt well survive and produce offspring that will survive. According to his theory, natural selection explains how diverse species could evolve from a common ancestor.

Declaration of Independence: Written in 1776 to declare America's freedom from British rule.

Devil and His Demons: They were once angels of God who rebelled and were cast out of Heaven. Now, they strive to bring hate, destruction, and evil into the world and the lives of people. They are sometimes referred to as evil spirits or the spiritual enemy of believers. Their eternal fate is the Lake of Fire, where they will never again be able to harass God's people.

Dinosaurs: Large reptiles that roamed the Earth. They were created by Jesus. They were loaded onto Noah's Ark. Many were small, the size of a cow. Larger ones were likely loaded when they were young. It is said that they went extinct after the flood, but the reason is uncertain. Perhaps the global ice age is the reason.

DNA: Most Deoxyribonucleic acid (DNA) is located in the nucleus of cells, while some is also found in the mitochondria (known as mtDNA). DNA is passed on to offspring. Therefore, it is used by people who want to trace their biological ancestors. The mtDNA in cells converts the energy from food into forms that the cells use.

DNA Breakthrough Finally Solves the Lost Colony Mystery — Genetic Evidence Reveals What Really Happened to Roanoke's Settlers: Description of how DNA helped solve this mystery.

Easter: Christian holiday celebrating the resurrection of Jesus Christ. Orthodox Easter takes place between April 4 and May 8, following the first full moon after Passover.

End Times: The seven years on Earth when the Anti-Christ attempts to rule in God's place. God destroys the power of the Anti-Christ, but the Earth is ravaged and unsuitable for life when it ends.

Evolution: The theory that all living creatures evolved over millions of years from ancient ancestors. The problem is that there are "missing links" that prevent a direct descent, an uninterrupted connection between the species. Charles Darwin is considered to be the father of the theory of evolution.

Faith: Biblical faith is informed faith. It is confidence and trust that God is who he says he is and will do what he says (Hebrews 11:1–3). Faith requires action; otherwise, it's false (James 2:26). Faith grows over time, deep within the soul, as people experience God and allow him to apply his living word to transform their hearts and lives. Belief in the theory of evolution is "blind faith" because there is no evidential substance to support it.

Flat Earth: This theory is considered a conspiracy aimed at preventing people worldwide from knowing that the Earth is round. I find that only people who like conspiracy theories believe in this.

Full Armor of God: Each piece of the armor of God must be worn in spiritual battle.

God (Father God): When we see the word "God" in the Bible, it typically refers to the first person of the Trinity. Sometimes, it also refers to the Trinity, emphasizing their oneness. In the Hebrew language of the Old Testament, God gives himself one personal name (Exodus 3:13–14). Because the people of Israel thought of this as his sacred name, they referred to it as YHWH, leaving out the vowels. This is sometimes written in English Bibles as I AM WHO I AM, lord (with small capital letters), Adonai, or Jehovah. Many names and titles for God reveal aspects of his divine nature and represent who he is, such as *El Shaddai* (God Almighty) or *Jehovah-shalom* (God is our peace).

Good News/Gospel: The Good News is that Jesus Christ died for the sins of every person, was buried, and was raised to life on the third day, according to the scriptures. People are born-again and become Christians only when they respond to this Good News by accepting Jesus into their hearts and lives as Savior and Lord.

Heaven: When the physical body of a follower of Jesus Christ dies, their spirit/soul goes immediately to Heaven. The Father God, Jesus Christ, myriads of angels, and other believers welcome them with loving and joyful arms. Heaven is the eternal abode of God (Father God, his Son, Jesus Christ, and the Holy Spirit), even though they exist everywhere. Since God is supernatural, it's supernatural as well. It's the place where God presides over the entire creation and his multitude of mighty angels. Heaven above comes down to become the new Jerusalem at the end of time, where all believers will exist eternally, enjoying God's presence on the new earth (Revelation 21:1–3).

Hell: When the physical body of those who reject Jesus Christ dies, their spirit/soul goes immediately to Hell. After the Great White Throne Judgment, they will be sent to exist eternally in the Lake of Fire, where they will never experience God's presence, love, and peace (Revelation 20:11–15).

Holy Spirit (Spirit of God and Christ): The Holy Spirit is identified as the third person of the Trinity. He has many names throughout the Bible, including the Spirit of God and Spirit of Christ. These indicate he is divine, fully God in nature, and part of the Trinity. The Holy Spirit opens people's hearts and minds to understand the Gospel that Jesus died for their sins, was buried, and resurrected to provide eternal life with God. When they profess their faith in Jesus Christ as Savior and ask him to be the Lord of their life, the Holy Spirit comes to live within them. Only then are they born-again and become a Christian.

Holy Week: The week leading up to Good Friday and Easter Sunday.

Hope: Hope, like faith, is an assurance that God will do what He says He will. There is a sense of confidence and certainty in biblical hope (Hebrews 11:1-3). Hope and faith are based on truth, not wishful thinking. Therefore, all hope in God comes from His Holy Spirit and can bring joy and peace as believers trust Him. Life without faith in Jesus Christ is a life without God and without hope (Ephesians 2:12).

Jamestown, Virginia: The English settled there in 1613.

Japan: The author had multiple experiences in Japan, including sumo wrestling.

Jesus Christ (Savior and Lord): He is identified as the second person of the Trinity. Jesus is his personal name (Matthew 1:21). Christ is a title meaning "the one chosen by God to save his people." He is fully God and fully divine in his nature. He is found throughout the entire Bible and has many names and titles that help us understand who he is and what he does.

Love Languages: Five ways people can express their love for a partner.

Millennial Reign of Christ: By the end of the seven-year Tribulation, the Earth will be completely destroyed. The Earth will not be able to sustain human or any other life. In Revelation 19 and 20, we read about the Millennial Reign of Christ on Earth, which follows this horrific period. We see that all the saved who were martyred by the antichrist and his armies will reign on Earth with Jesus for a thousand years.

Miracles: Something that defies the natural order of life.

Mystery: Some verses and topics in the Bible are simply a mystery. A mystery in the Bible is something for which God has partially concealed its meaning or even hidden it entirely.

Myth: These occur everywhere in a society. However, there are no myths included in the Bible. For example, many myths about Heaven and Hell are not in the Bible.

As such, they are untrue and should be disregarded by Christians.

New Covenant (Contract, Agreement): The Old Covenant pointed to the need for the New Covenant, based on grace and faith rather than on people's efforts. The purpose of the New Covenant is to offer salvation as the free gift of God's grace (unmerited favor). Jesus Christ is the only one who died and rose again to fulfill the legal requirement of the Old Covenant to bring the free gift of salvation and eternal life to everyone who would accept Him as their Savior and Lord.

New Heaven and New Earth. God will ultimately terminate this current creation by fire because of the rampant rebellion of people against him. He has delayed doing so because he wants every person to repent and be saved. After its destruction, he will create an entirely new heaven and a new Earth. This will be a wonderful place of eternal existence for the followers of Jesus Christ. The Father and Jesus Christ will come down from Heaven to live among them in this new creation. There will never again be pain, suffering, or sin in this eternal home with God. There will be no crime or murder because no one without Jesus Christ will be there.

Passover: God commanded the Jewish nation to celebrate Passover in remembrance of His passing over them in Egypt.

Passover Seder: The meal for the Passover celebration.

Prophecy: A message from God through the Holy Spirit to inform, guide, or warn people.

Radiometric Dating: Scientists, especially geologists, use radiometric dating to estimate the age of rocks and the fossils they contain. This is also called radioisotope dating or radioactive dating.

Sabbath: Fourth of the Ten Commandments, which is not required of Christians to please God.

Sin: No one is perfect. No one can be good enough to earn salvation. Everyone does what displeases God. However, only one sin separates people from eternal life with God. It's the sin of not accepting Jesus Christ as Savior and Lord.

Sovereignty of God: The Triune God is in complete control of everything, everywhere, all the time. Nothing exists outside of His absolute authority. He allows people to exercise their free will within the limits of His sovereign will.

Spirit, Soul, and Body: From the Bible, we see that people have a spirit, soul, and body. This is their threefold human nature.

Supernatural World: There is something beyond the physical, natural world. It's the supernatural world. It's the unseen world around us. The Triune God and His angels are supernatural, and everything they do is supernatural. The devil and his demons are supernatural, also, and everything they do is supernatural.

Ten Commandments: These were among the 613 commandments God gave to the nation of Israel (Deuteronomy 5:6-22). They are a short list of concise statements about how the people of Israel were to interact with God and with other people.

Trinity (Triune God): The terms "Trinity" and "Triune God" don't appear in the Bible. Biblical scholars created them to help people understand the unity, yet separate reality, of the three persons of God. They are God the Father, his Son, Jesus Christ, and the Holy Spirit. They are each equally God in every aspect of their being and nature.

Truth: It's an absolute that doesn't change with the culture or social norms. No lies, deceptions, or half-truths are involved with biblical truth. Everything in the Bible is true because it's God-breathed (2 Timothy 3:16).

Universe: In Genesis 1:1, we see the creation of the first universe, and in Revelation 21:1, we see the uniquely created new universe.

Bibliography

Berlin, Irving. *God Bless America. Patriotic Song*. Written in 1918 and revised in 1938. Wikipedia. Accessed 7/7/2026. *https://en.wikipedia.org/wiki/God_Bless_America.*

Buckley, Jay H. Britannica Editors. "Louis and Clark Expedition." Accessed 5/7/2026. *https://www.britannica.com/event/Lewis-and-Clark-Expedition.*

———. Britannica Editors. *San Francisco earthquake of 1906*. Published April 11, 2026, History." Accessed 4/18/2026.

———. Britannica Editors. *Berlin Wall, Berlin, Germany*. Updated March 17, 2026. Accessed 4/17/2026.

———. Britannica Editors. Ray, Michael. *In What Order Were the 13 American Colonies Established?* Accessed 5/7/2026.

———. Britannica Editors. *United States History, Louisiana Purchase*. Published April 25, 2026. Accessed 5/7/2026.

Bryan, Paul. *Who Is This God? A Handbook for Life with Him*. Houston: Lucid, 2023.

———. *Unraveling Creation and Evolution through Science and the Bible*. Eugene: Resource Publications.

Christian History. *Debating Jesus's Divinity, Revisiting the Council of Nicaea and its 1700th Anniversary*. Issue 58, Christian History Institute: Worcester, 2026.

———. Christian History. *Taking Care of Church Business*. Issue 58, Christian History Institute: Worcester, 2026.

Davis, Tchiki. *What Are the 5 Love Languages? Definition and Examples." "How do you receive love?* Gary Chapman's love languages can help you find out. Accessed 4/2/2026. *https://www.psychologytoday.com/us/blog/click-here-happiness/202009/what-are-the-5-love-languages-definition-and-examples.*

DNA. Cleveland Clinic. Last Updated 7/9/2025. Accessed 4/4/2026. *https://my.clevelandclinic.org/health/body/dna*

Dobrijevic, Daisy with contributions from Stefanie Waldek. *Northern lights (aurora borealis): What they are and how to see them*. Updated May 6,

2025. Accessed 4/12/2026. *https://www.space.com/15139-northern-lights-auroras-earth-facts-sdcmp.html.*

Downtown San Francisco. *120 Years Later: Remembering the 1906 San Francisco Earthquake and Fire. https://downtownsf.org/post/120-years-later-remembering-the-1906-san-francisco-earthquake-and-fire.*

Glanze, Patrick. *12 Fascinating Facts About the Human Brain Scientists Have Discovered Through Neuroscience.* Published by the Science Times, Feb 11 2026, 07:40 AM EST.

Historian. *Fun Facts, 10 Facts About the Phoenix.*" Published July 20, 2023. Accessed 4/18/2026.

July 20, 2023 by The Historian.

History.com Editors. "Christopher Columbus." https://www.history.com/articles/christopher-columbus.

History.com Staff. *What Happened to the 'Lost Colony' of Roanoke? How could 115 people just vanish?* Accessed 4/2/2026. *https://www.history.com/articles/what-happened-to-the-lost-colony-of-roanoke.*

Hustrulid, William Andrew. Brittanica Editors. *Placer Mining.* Accessed 4/14/2026.

The Joy of English. *Half of the English language is of French origin.* Published October 26, 2023. Accessed 4/2/2026. *https://readable.com/blog/half-of-the-english-language-is-of-french-origin.*

Kennedy, Cain. *Roman Concrete Vs. Cement: Unraveling The Ancient Building Mystery.* Last Updated January 8, 2026. Accessed 4/10/2026.

Kinney, Peter. *The Silent Majesty of Mesoamerican Pyramids.* Posted April 7, 2026. Accessed 4/10.2026. *https://www.factinate.com/things/silent-majesty-mesoamerican-pyramids.*

Margaritoff, Marco. Edited By Jaclyn Anglis. *What Is a Kraken, The Frightening Sea Monster From Scandinavian Lore?* Published May 13, 2024. Accessed 5/2/2026. *https://allthatsinteresting.com/kraken*

Passover: Jewish Virtual Library. Accessed 4/1/26. *https://jewishvirtuallibrary.org/passover-history-and-overview*

Passover Seder: Lerner, Barry, Rabbi. *My Jewish Learning. https://www.myjewishlearning.com/article/how-to-conduct-a-seder/*

Staff Contributor. *DNA Breakthrough Finally Solves the Lost Colony Mystery — Genetic Evidence Reveals What Really Happened to Roanoke's Settlers.* Written October 6, 2025. Accessed 4/2/2026.

Wanderling. *The Grand Canyon, lost Aztec gold, Spanish conquistadors, and the Colorado River.*" Accessed 4/11/2026. https://the-wanderling.com/hunt.html.

Whitelaw, Henry. *The Scottish Colonization of Nova Scotia.* Accessed 4/5/2026. *https://www.historic-uk.com/HistoryUK/HistoryofScotland/The-Colonisation-of-Nova-Scotia.*

Wikipedia. *Biblical Canon.* Accessed 3/23/2026. *https://en.wikipedia.org/wiki/Biblical_canon.*

———. *Pocahontas.* Accessed 4/2/2026. *https://en.wikipedia.org/wiki/Pocahontas*

www.ingramcontent.com/pod-product-compliance
Lightning Source LLC
LaVergne TN
LVHW022052200926
848481LV00003B/873

* 9 7 9 8 3 8 5 2 8 8 0 2 1 *